PolyMath Assessments: Kindergarten

Honora Wall, M.A., Applied Curriculum & Instruction

PolyMath Assessments: Kindergarten

Honora Wall, M.A., Applied Curriculum & Instruction

PolyMath Publishing, Ocala, Florida
©2018, What the (f)unction, Inc.
Illustrations © Chris Wall, Badly Drawn Robot

ISBN: 978-1-7327601-1-0

PolyMath Assessments: Helping all students succeed

Welcome to PolyMath

During my many years working as an educator, I have taught 1st grade, 2nd grade, middle school math, Pre-Algebra, Algebra I, and Geometry. As a private tutor, I've helped students understand and master math from Kindergarten to College Algebra, including SAT and ACT math. In each of these settings, I worked with students who had some sort of trouble "getting" math. Helping these learners led me to focus on curriculum and instruction for both my Master's and Doctorate degrees. I asked myself, how can we affect learning by changing the way we teach and the formats we use to assess mastery?

The answer is in PolyMath. Poly means *many*. We need many methods of instruction, many types of curriculum, and many means of assessing learning, because there are so many different reasons why a person has difficulty with math. These assessments use the most current research and recommended best practices for students with dyscalculia, dysgraphia, high-functioning autism, ADHD/ADD, processing disorders, and other physical, emotional, and neurological conditions. PolyMath Assessments focus on measuring ability and mastering skills by reducing graphics, text, and other distractions. Assessments could be oral, matching, or true / false, and accommodations are included when appropriate. While PolyMath Assessments can be used by any student, these workbooks were written specifically for students with dyscalculia, dyslexia, dysgraphia, high-functioning Autism, processing disorders, or severe ADHD/ADD. With their reduced text and graphics and fewer distractions on a page, they are also a great resource for ESOL/ELL students. The assessments are drawn directly from state standards and each one includes accommodations designed to let these students showcase their abilities.

PolyMath Assessments are written for parents and teachers who want to support students where they are. They allow students to demonstrate knowledge without the extra graphics, extra problems, or extra topics found in other assessments. They use appropriate accommodations that address the many reasons why math is hard for some students. These accommodations include more white space on the page, fewer problems written on a page (in this case, an assessment may be spread over two or three pages), an included number line or word bank, or graphics that support the problem. These are not modifications; they do not make the work any easier. They simply provide the research-proven design components that help students focus.

Each page has space for your state standard. The standards are listed in the index table. Simply go to the index, look up the assessment by page number, and find your state on the table. Write your standard on the assessment page in the space provided. At the end of the year, you will have a record of mastery for each state standard.

Record keeping for home school students

Home schooling parents are required to keep a portfolio of work for their students, and many states require a certified teacher to sign off on annual progress. PolyMath Assessments help parents save time and space by creating the math portfolio for you! We've compiled all the state standards for every grade level; teachers can easily see that your child has passed every required assessment. The topic, plus the standard itself, is clearly marked on each page. You can use the curriculum of your choice while maintaining records easily with PolyMath Assessments.

As a home school parent, you have the choice of using any curriculum that works best for your child and your family. While you have many great options to choose from, not every option makes it easy for a certified teacher to see whether or not your child has reached all of the required state standards they need to move on to the next grade. PolyMath Assessments makes it easy for you and for them—simply use the curriculum of your choice for daily math instruction and practice, and when your child is ready, have them take the appropriate assessments in this book. At the end of the year, you've got all your math assessments together, plus examples of the curriculum you used. The annual report will be easy to complete.

Assessing students with specific learning requirements

The Individuals with Disabilities Education Act (IDEA) protects children and families by making sure all students receive an equitable education regardless of differences in physical or neurological abilities. IDEA also protects accommodations for students. As parents, and as teachers, it can be difficult to know which accommodations should be used for which student. PolyMath Assessments have taken the guesswork out of this equation by including them in the design of each assessment. Only a licensed, trained professional can determine if a child has a learning disability or a medical condition that impedes their education. If you feel your child or student is struggling more than others, or if you see that they are falling behind other children, please do not wait to have them evaluated. The earlier a child receives interventions and instruction appropriate to their needs, the greater success they can achieve.

Good news for teachers

Teachers, I know from my own experience that your job is more difficult than anyone outside the field can imagine. I know you have students with many different abilities in your class at any given time. I know you can't reasonably adjust instruction AND assessments for all students, align changes with current standards, keep up with research, and meet reporting requirements. It is just an overwhelming amount of work. I also know that you *want* to do all of these things, because I know how dedicated you are to your students. My hope is that PolyMath Assessments can offer you some help—assessments with accommodations, proven to work, that are aligned to your state standards. I hope these will reduce your workload while meeting your students' needs and increasing what you can offer the children in your classroom.

Table of Contents

Page 6	....	Count to 100 by ones and by tens
Page 8	....	Count forward by ones from a given number
Page 10	....	Read and write numerals from 0 to 20
Page 16	....	Understand numbers and quantities
Page 20	....	Count objects in a group, up to 10
Page 22	....	Compare two groups by matching or counting
Page 26	....	Compare two numbers between 1 and 10
Page 28	....	Addition and subtraction word problems
Page 30	....	Decompose numbers into addition pairs
Page 32	....	Use objects or drawings to make ten
Page 36	....	Add and subtract within five
Page 38	....	Decompose 11 through 19 by tens and ones
Page 40	....	Weight, length, and height
Page 42	....	Taller and shorter
Page 44	....	Classify and count with a Venn Diagram
Page 46	....	Describe the Location
Page 48	....	Naming shapes
Page 50	....	2-D and 3-D shapes
Page 52	....	Same and different: Shapes
Page 54	....	Drawing Shapes
Page 56	....	Use Shapes to Make New Shapes
Page 58	....	Resources
Page 65	....	Answer Keys
Page 92	....	Glossary
Page 94	....	Index of State Standards

Kindergarten 1   Count to 100 by ones and by tens   Standard:_____

This is a verbal assessment. Students may need to recite numbers in smaller chunks, rather than counting 1 to 100 at one time. Use this chart to record the dates of the assessment, the starting number, and the last number said before the student miscounts. Continue the assessment the next day, starting at the last number of the previous day, until the student counts to 100.

Student: _____

Date	Starting Number	Ending Number	Teacher Initials

© PolyMath Publishing 2019   May not be reproduced without permission

Kindergarten 2   Count forward by ones from a given number   Standard:_____

This is a verbal assessment. Record the date, the number you ask the student to start with, and the student's response. For example, ask the student to start at the number 3, and count 4 more than that. Note any tools used (counting on fingers, number line, 1-25 chart, etc.).

Student: _____

Date	Starting Number	Ending Number	Teacher Initials

Tools used: _____

Kindergarten 3   Read and write numerals from 0 to 20   Standard:_____

<u>Part 1 of 3:</u> This is an oral assessment. Read the numbers 0-20 to the student, in no particular order. Ask the student to write down the number you say. *Note: The student should write the cardinal number (i.e., 4), not word form (i.e., four). Students with writing issues such as dysgraphia may show the teacher a number card instead of writing the number. In this case, write down the number shown and whether or not it was correct.*

Student: _____   _____ / 20 = _____ %

Kindergarten 3   Read and write numerals from 0 to 20   Standard:_____

Part 2 of 3: This is an oral assessment. Read the numbers 11-20 to the student, in no particular order. Ask the student to write down the number you say. *Note: The student should write the cardinal number (i.e., 4), not word form (i.e., four). Students with writing issues such as dysgraphia may show the teacher a number card instead of writing the number. In this case, write down the number shown and whether or not it was correct.*

Student: _____   _____ / 10 = _____ %

Kindergarten 3   Read and write numerals from 0 to 20   Standard: _____

Part 3 of 3: This is an oral assessment. Have the student read each number to you. Circle any number the student says incorrectly.

Student: _____   _____ / 20 = _____ %

3	12	7	5	11
20	8	19	17	4
1	14	18	9	10
2	6	16	13	15

© PolyMath Publishing 2019   May not be reproduced without permission

Kindergarten   4   Understand numbers and quantities   Standard:_____

Part 1 of 2: Write the number that shows how many objects are in the picture.

Student:_____   _____/ 11 = _____%

	☆ ☆ ☆
	☆☆☆☆☆ ☆☆☆☆☆
	△△△ △△△ △△△ △△△ △△ △△ △△
	☆ ☆ ☆ ☆ ☆
	△△△ △△△ △ △△ △△
	☆ ☆
	○○○ ○○○ ○○○ ○○ ○○ ○○
	☆
	☆ ☆ ☆ ☆ ☆ ☆
	○○○ ○○○ ○○○ ○○○ ○○ ○○ ○○ ○○

© PolyMath Publishing 2019   May not be reproduced without permission

Kindergarten   4   Understand numbers and quantities   Standard:_____

Part 2 of 2: Write the number that shows how many objects are in the picture.

Student:_____   _____/ 10 = _____ %

	○ ○ ○ ○ ○ ○
	□ □ □ □ □ □ □ □ □ □ □
	✿ ✿ ✿ ✿
	☆ ☆ ☆ ☆ ☆ ☆ ☆ ☆ ☆ ☆
	○ ○ ○ ○ ○ ○ ○ ○
	✿ ✿ ✿ ✿ ✿ ✿ ✿ ✿ ✿ ✿ ✿ ✿ ✿ ✿
	△ △ △ △ △ △ △ △ △ △ △ △ △ △ △ △ △ △ △
	○ ○ ○ ○ ○ ○ ○ ○ ○ ○ ○ ○ ○ ○
	☆ ☆ ☆ ☆ ☆ ☆ ☆ ☆ ☆
	△ △ △ △ △ △ △ △ △ △ △ △ △ △ △ △

© PolyMath Publishing 2019   May not be reproduced without permission

Kindergarten   5   Count objects in a group, up to 10   Standard: _____

Tell the student a number from 1 to 10. Have the child draw a group of that many objects, or they can make a group out of classroom objects. Drawings can be of any shape the child is comfortable making.

Student: _____   _____ / 10 = _____ %

Number: _____   Objects: _____

Number: _____   Objects: _____

Number: _____   Objects: _____

Number: _____   Objects: _____

Number: _____   Objects: _____

Number: _____   Objects: _____

Number: _____   Objects: _____

Number: _____   Objects: _____

Number: _____   Objects: _____

Number: _____   Objects: _____

© PolyMath Publishing 2019   May not be reproduced without permission

Kindergarten 6   Compare two groups by matching or counting   Standard: _____

Page 1 of 2: In each box, circle the group that has MORE. These two pages may be given to the student at the same time; the assessment is separated to reduce eye strain and visual confusion for the student.

Student: _____   _____ / 5 = _____ %

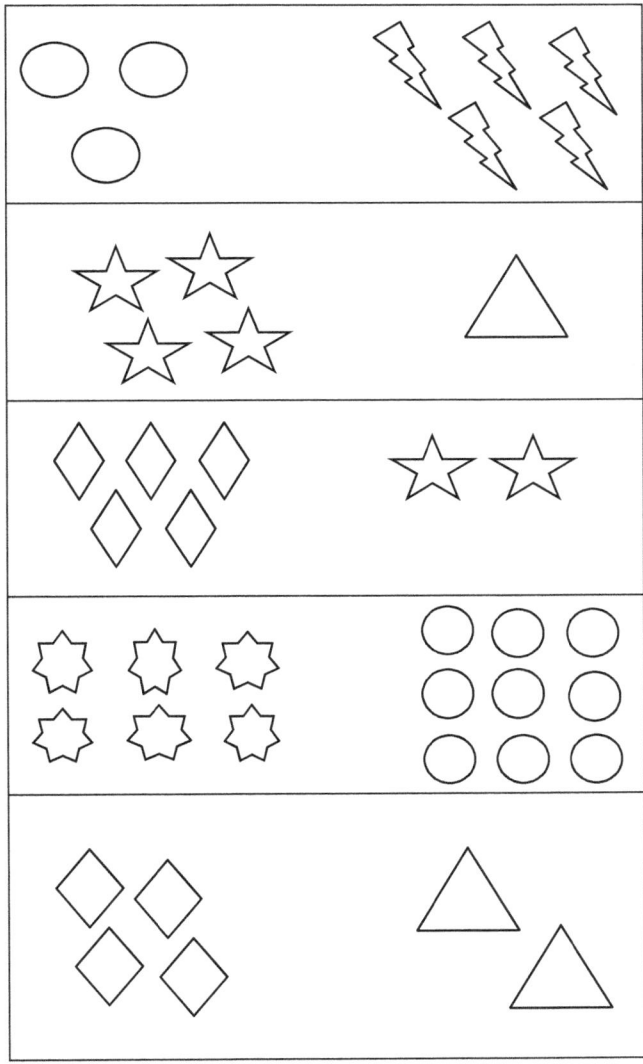

© PolyMath Publishing 2019   May not be reproduced without permission

Kindergarten   6   Compare two groups by matching or counting   Standard:_____

Page 2 of 2: In each box, circle the group that has MORE. These two pages may be given to the student at the same time; the assessment is separated to reduce eye strain and visual confusion for the student.

Student: _____   _____ / 5 = _____ %

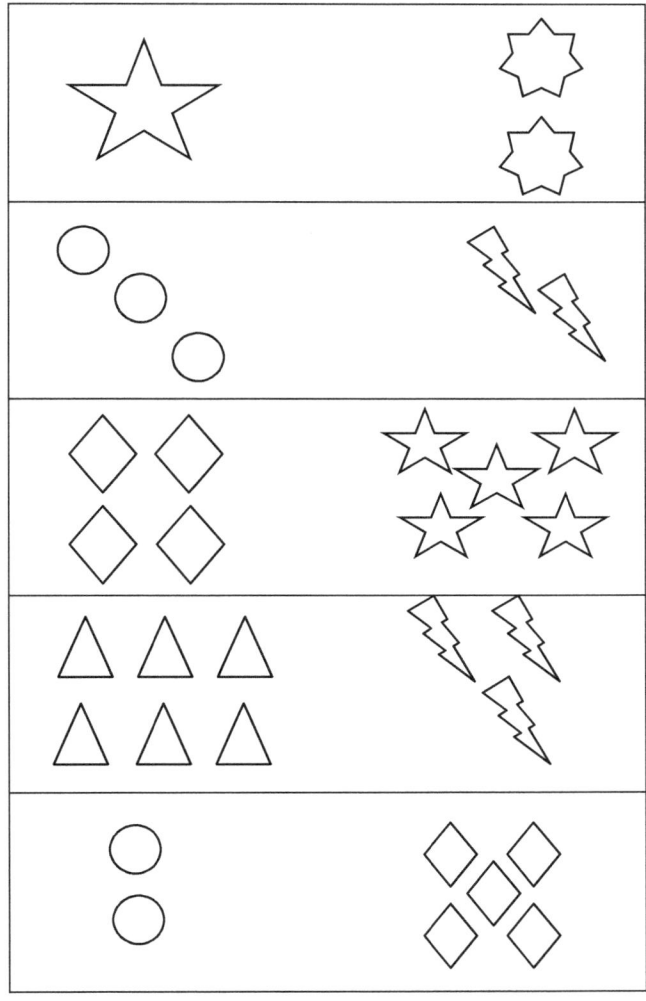

© PolyMath Publishing 2019   May not be reproduced without permission

Kindergarten   7   Compare two numbers between 1 and 10   Standard:_____

In each box, circle the bigger number.

Student:_____   _____/ 12 = _____ %

3 10	5 0	10 4
6 2	1 5	8 9
0 7	4 2	10 6
8 3	6 7	1 9

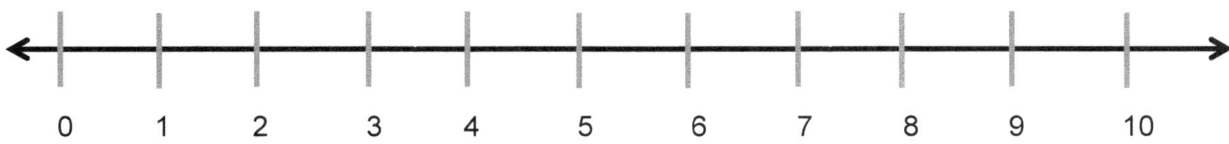

© PolyMath Publishing 2019   May not be reproduced without permission

Kindergarten   8   Addition and subtraction word problems   Standard:_____

Read each problem to the student. Have them decide if they should add or subtract to solve. Grade one point for the operation (add or subtract) and one point for the answer.

Student: _____   _____ / 10 = _____ %

1. James has three apples and Sam has two apples.
   How many apples do they have all together?   _____

2. Luis has three pencils. Nancy has five pencils.
   If they put all their pencils in one basket, how
   many pencils are in the basket?   _____

3. Sarah has two goldfish. Alex has four goldfish.
   How many fish do they have together?   _____

   Draw Sarah's goldfish here:
   Draw Alex's goldfish here:

4. Tommy has ten crayons and gives five
   crayons to Sharon. How many crayons are left?   _____

   Draw Tommy's crayons here:

5. If Maria has six apples, and shares two apples with
   her friend Max, how many apples does she have left?   _____

© PolyMath Publishing 2019   May not be reproduced without permission

Kindergarten   9   Decompose numbers into addition pairs   Standard:_____

Match the number to the addition pair that makes an equal amount. Students may draw a matching line or color matching amounts.

Student:_____   _____ / 10 = _____ %

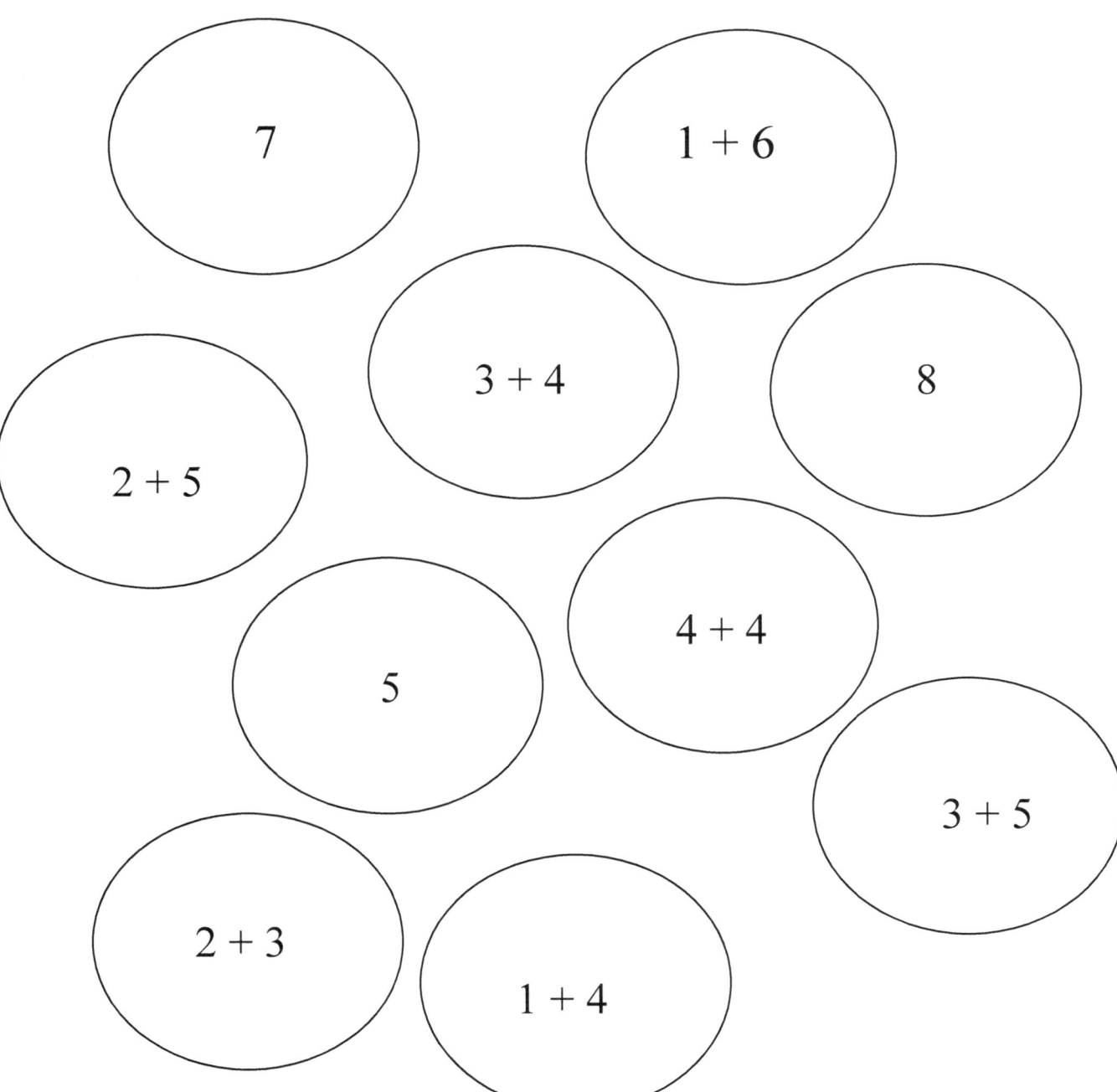

© PolyMath Publishing 2019   May not be reproduced without permission

Kindergarten   10   Use objects or drawings to make ten   Standard: _____

Page 1 of 2: Draw a line connecting the boxes to make a set of ten. These two pages may be given to the student at the same time; the assessment is separated to reduce eye strain and visual confusion for the student.

Student: _____   _____ / 5 = _____ %

© PolyMath Publishing 2019   May not be reproduced without permission

Kindergarten   10   Use objects or drawings to make ten   Standard: _____

Page 2 of 2: Draw a line connecting the boxes to make a set of ten. These two pages may be given to the student at the same time; the assessment is separated to reduce eye strain and visual confusion for the student.

Student: _____   _____ / 5 = _____ %

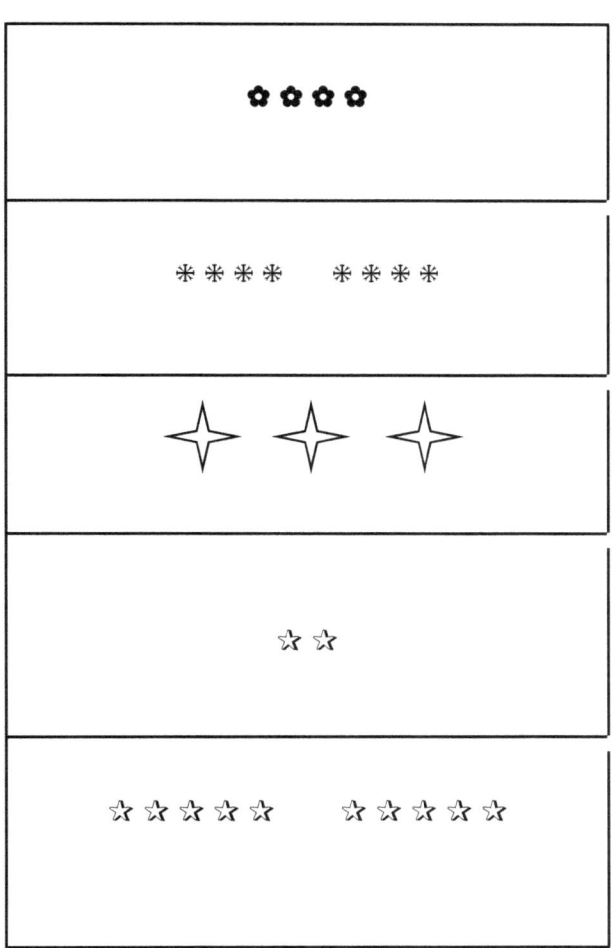
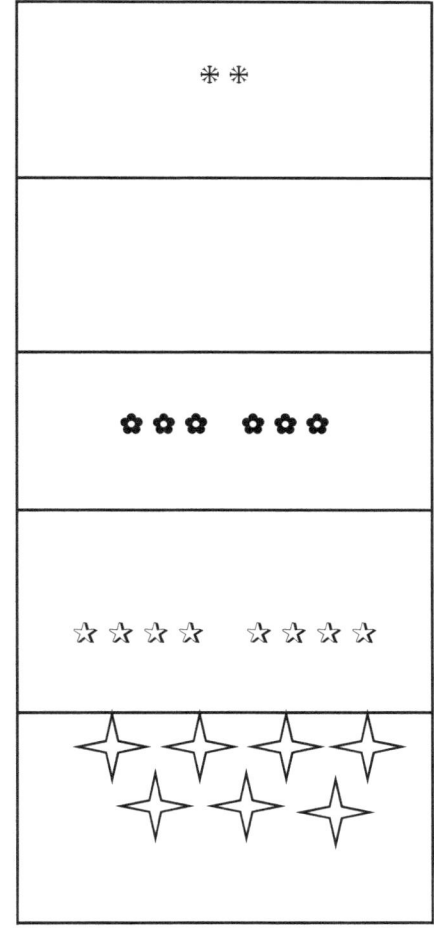

© PolyMath Publishing 2019   May not be reproduced without permission

Kindergarten   11   Add and subtract within five   Standard:_____

Students with dyscalculia may use support materials, such as a 1-20 chart, number line, or counting on their fingers.

Student: _____   _____ / 20 = _____ %

2 +3	4 +1	3 +3	4 +0
5 -3	3 -1	4 -3	5 -0
3 +1	2 +1	2 +2	1 +0
2 -1	4 -1	3 -3	4 -0
1 +3	5 -0	2 +2	3 -0

0 1 2 3 4 5 6 7 8 9 10

Kindergarten   12   Decompose 11 through 19 by tens and ones   Standard: _____

Write the number in tens and ones, like this: 12 = 10 + 2.

Student: _____   _____ / 10 = _____ %

11	
15	
18	
13	
12	
16	
19	
14	
17	
20	

© PolyMath Publishing 2019   May not be reproduced without permission

Kindergarten 13   Weight, length, and height   Standard: _____

This is a verbal assessment. Begin with any object in the classroom and ask the student the following series of questions. Record the student's choices with a "yes" or "no" response and a description of the new object.

Student: _____   _____ / 10 = _____ %

Initial object shown to the student: _____

1. Can you find something that is longer than this? _____

2. Can you find something that is shorter than this? _____

3. Can you find something that is taller than this? _____

4. Can you find something that is heavier than this? _____

5. Can you find something that is lighter than this? _____

6. Can you find something that is wider than this? _____

7. Can you find something that is thinner than this? _____

8. Can you find something that is the same height as this? _____

9. Can you find something that is the same weight as this? _____

10. Can you find something that is the same length as this? _____

© PolyMath Publishing 2019   May not be reproduced without permission

Kindergarten 14	Taller and shorter	Standard: _____

Student: _____	_____ / 10 = _____ %

Circle the object that is TALLER	Circle the object that is SHORTER
A A	5 5
◇ ◇	✸ ✸
○ ○	z X
9 2	◇ ◇
✸ ✸	○ ○

Kindergarten   15   Classify and count with a Venn Diagram   Standard: _____

Use the table to fill in the graph.

Student: _____   _____ / 4 = _____ %

Mr. Johnson's class pets:

	Dog	Cat
Jimmy	Yes - Sparky	No
Sam	Yes- Bloo	Yes- RC
Tammy	No	Yes- Fluffy
John	Yes- Jack	Yes- Jill
Lizette	No	No

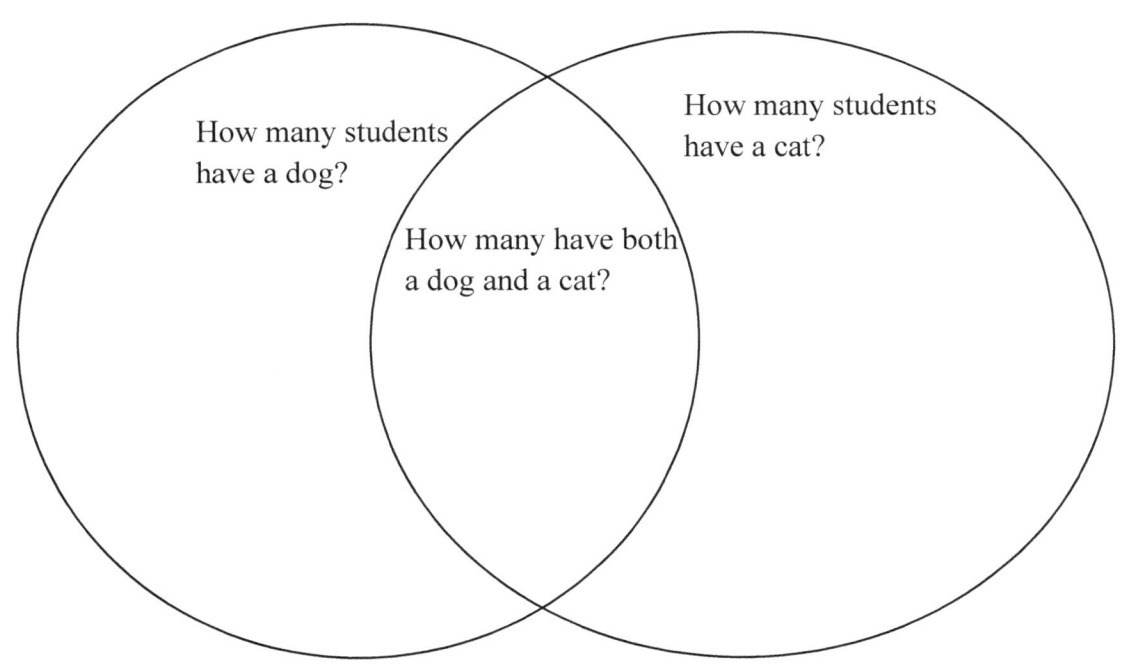

How many students have no pets? _____

Kindergarten 16	Describe the Location	Standard:_____

Circle the correct answer to complete the sentence.

Student:_____	_____ / 10 = _____ %

The cats are   INSIDE / UNDER the box.

The picture is NEXT TO / ABOVE the table.

The tree is to the RIGHT / LEFT of the rug.

The table is ON / NEXT TO the rug.

The box is UNDER / ON the table.

The rug is UNDER / ON the table.

The picture is to the LEFT / RIGHT of the tree.

The tree is BEHIND / IN FRONT OF the table.

The rug is BEHIND / IN FRONT OF the tree.

The cats are to the RIGHT / LEFT of the tree.

© PolyMath Publishing 2019   May not be reproduced without permission

Kindergarten 17			Naming shapes			Standard: _____

Draw a line matching the word to its shape. Then color all the triangles blue, all the circles red, all the squares green, and all the rectangles yellow.

Student: _____   _____ / 12 = _____ %

Triangle

Circle

Square

Rectangle

© PolyMath Publishing 2019					May not be reproduced without permission

Kindergarten   18   2-D and 3-D shapes   Standard: _____

This is an oral assessment. The student can bring an object to you, or they can point out an appropriate object in the room (for example, they might point to a clock or a globe, but not carry one to you). Record the objects the student uses for each shape.

Student: _____   _____ / 8 = _____ %

Can you find a circle? _____

Can you find a triangle? _____

Can you find a square? _____

Can you find a sphere? _____

Can you find a cube? _____

Can you find a prism? _____

Can you find a rectangle? _____

Can you find a pyramid? _____

© PolyMath Publishing 2019   May not be reproduced without permission

Kindergarten   19   Same and different: Shapes   Standard:_____

This can be a written or an oral assessment. Give the student a variety of objects: circles, triangles, squares, rectangles, cubes, or any other shape. Let the student describe the ways shapes are similar and different. Ask them about the size, number of corners, flat or solid, or color of the shapes. See if the student can find ten ways they can describe the objects as similar and different.

Student: _____   _____ / 10 = _____ %

Kindergarten   20   Drawing Shapes   Standard: _____

Draw each shape and answer the questions about the shape. Teachers can write in student answers to the questions. Drawing the shape counts as one point, and each question counts as one point.

Student: _____   _____ / 9 = _____ %

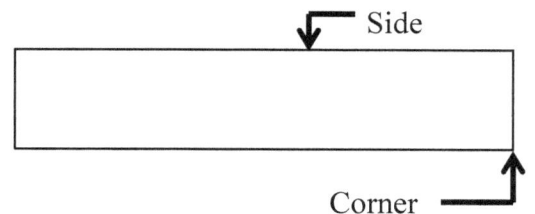

Example: This is a <u>rectangle</u>. It has <u>four</u> sides. It has <u>four</u> corners.

A triangle has _____ sides.

A triangle has _____ corners.

Draw a triangle.

A square has _____ sides.

A square has _____ corners.

Draw a square.

A circle has _____ sides.

A circle has _____ corners.

Draw a circle.

© PolyMath Publishing 2019   May not be reproduced without permission

Kindergarten   21   Use shapes to make new shapes   Standard: _____

Have the student use shapes such as triangles, squares, and rectangles to create new shapes. For example, two triangles can make a square or a rectangle. A rectangle and triangle can make a building. The first five shapes are required shapes, and the next five shapes are open to the child's imagination. The student can draw the shapes or use materials and objects in the classroom. Make a note of the materials used.

Student: _____   _____ / 10 = _____ %

Shape: _____ Materials used (drawing, paper cut-outs, objects):

Make a square out of two triangles. _____

Make a rectangle from two triangles. _____

Make a rectangle from two squares. _____

Make a building from squares and triangles. _____

Make a building from rectangles and triangles. _____

Student choice: _____

Student choice: _____

Student choice: _____

Student choice: _____

Student choice: _____

© PolyMath Publishing 2019   May not be reproduced without permission

1-20 Chart

Students who are overwhelmed by the amount of information in a standard 1-100 chart can use this chart for counting, adding, and subtracting.

1	2	3	4	5
6	7	8	9	10
11	12	13	14	15
16	17	18	19	20

21-40 Chart

Students who are overwhelmed by the amount of information in a standard 1-100 chart can use this chart for counting, adding, and subtracting.

21	22	23	24	25
26	27	28	29	30
31	32	33	34	35
36	37	38	39	40

© PolyMath Publishing 2019   May not be reproduced without permission

41-60 Chart

Students who are overwhelmed by the amount of information in a standard 1-100 chart can use this chart for counting, adding, and subtracting.

41	42	43	44	45
46	47	48	49	50
51	52	53	54	55
56	57	58	59	60

61-80 Chart

Students who are overwhelmed by the amount of information in a standard 1-100 chart can use this chart for counting, adding, and subtracting.

61	62	63	64	65
66	67	68	69	70
71	72	73	74	75
76	77	78	79	80

81-100 Chart

Students who are overwhelmed by the amount of information in a standard 1-100 chart can use this chart for counting, adding, and subtracting.

81	82	83	84	85
86	87	88	89	90
91	92	93	94	95
96	97	98	99	100

© PolyMath Publishing 2019   May not be reproduced without permission

Students may prefer this number line for counting, adding, and subtracting.

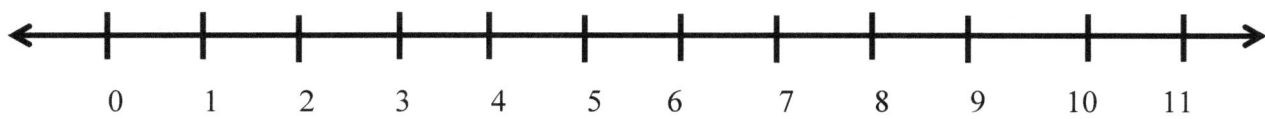

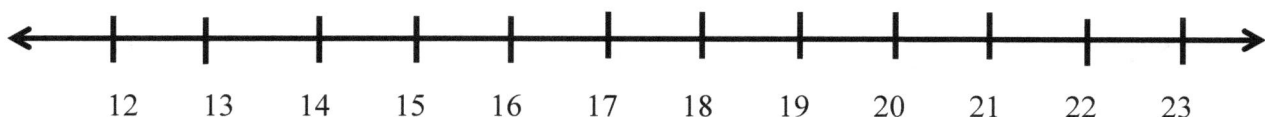

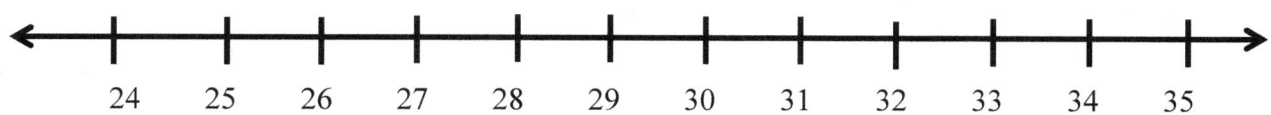

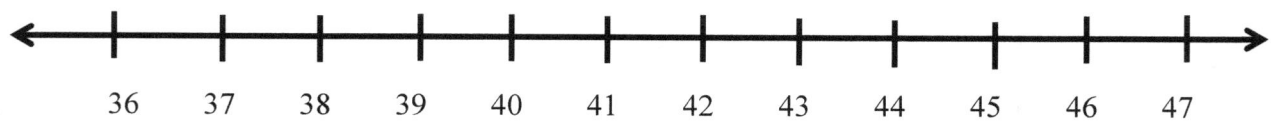

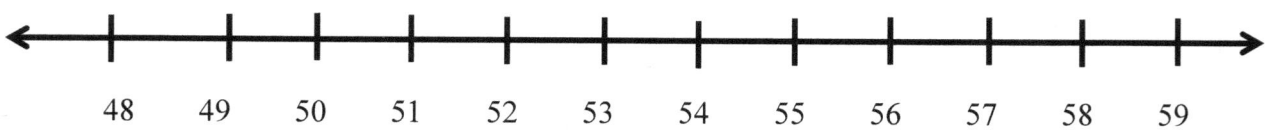

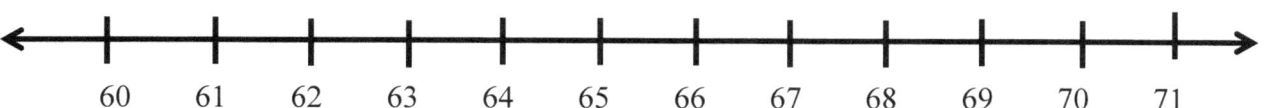

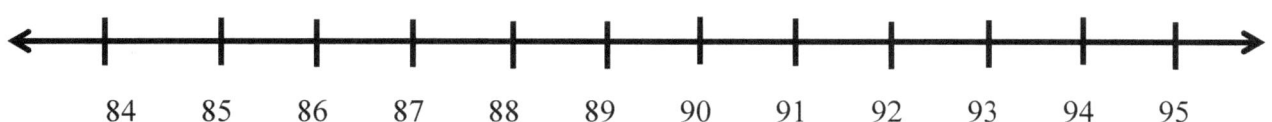

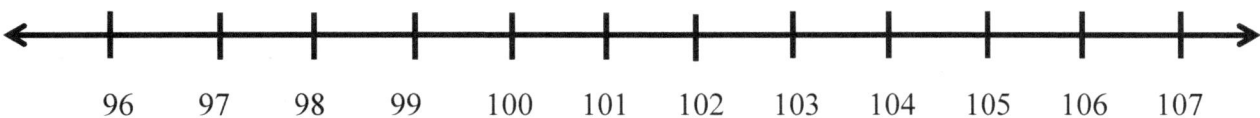

Kindergarten 1 Count to 100 by ones and by tens ANSWER KEY

This is a verbal assessment. Students may need to recite numbers in smaller chunks, rather than counting 1 to 100 at one time. Use this chart to record the dates of the assessment, the starting number, and the last number said before the student miscounts. Continue the assessment the next day, starting at the last number of the previous day, until the student counts to 100.

Student: _____ANSWER KEY_____

Date	Starting Number	Ending Number	Teacher Initials

To score, record the highest number the student counts to. Divide that number by 100. For example, if the student counts correctly up to 86 but makes repeated errors after that, even at the end of the school year, then 86/100 = 0.86 or 86%.

© PolyMath Publishing 2019   May not be reproduced without permission

Kindergarten   2   Count forward by ones from a given number   ANSWER KEY

This is a verbal assessment. Record the date, the number you ask the student to start with, and the student's response. For example, ask the student to start at the number 3, and count 4 more than that. Note any tools used (counting on fingers, number line, 1-25 chart, etc.).

Student: _____ANSWER KEY_____

Date	Starting Number	Ending Number	Teacher Initials

To score, record the highest number the student counts to. Divide that number by 100. For example, if the student counts correctly up to 86 but makes repeated errors after that, even at the end of the school year, then 86/100 = 0.86 or 86%.

Tools used: _____

© PolyMath Publishing 2019   May not be reproduced without permission

Kindergarten 3   Read and write numerals from 0 to 20   ANSWER KEY

<u>Part 1 of 3:</u> This is an oral assessment. Read the numbers 0-20 to the student, in no particular order. Ask the student to write down the number you say. *Note: The student should write the cardinal number (i.e., 4), not word form (i.e., four). Students with writing issues such as dysgraphia may show the teacher a number card instead of writing the number. In this case, write down the number shown and whether or not it was correct.*

Student: _____ ANSWER KEY _____ _____ / 21 = _____ %

The student should write the correct number. Backwards writing is acceptable for this assessment. To score, divide the number of correct answers by 21.

© PolyMath Publishing 2019   May not be reproduced without permission

Kindergarten 3   Read and write numerals from 0 to 20   ANSWER KEY

Part 2 of 3: This is an oral assessment. Read the numbers 11-20 to the student, in no particular order. Ask the student to write down the number you say. *Note: The student should write the cardinal number (i.e., 4), not word form (i.e., four). Students with writing issues such as dysgraphia may show the teacher a number card instead of writing the number. In this case, write down the number shown and whether or not it was correct.*

Student: _____ ANSWER KEY _____   _____ / 10 = _____ %

The student should write the correct number. Backwards writing is acceptable for this assessment. To score, divide the number of correct answers by 10.

Kindergarten 3   Read and write numerals from 0 to 20   ANSWER KEY

Part 3 of 3: This is an oral assessment. Have the student read each number to you. Circle any number the student says incorrectly.

Student: _____ ANSWER KEY _____   _____ / 20 = _____ %

3	12	7	5	11
20	8	19	17	4
1	14	18	9	10
2	6	16	13	15

Kindergarten   4   Understand the relationship between numbers and quantities   ANSWER KEY

Part 1 of 2: Write the number that shows how many objects are in the picture.

Student: _____ ANSWER KEY _____ _____ / 11 = _____ %

3	☆ ☆ ☆
10	☆☆☆☆☆ ☆☆☆☆☆
18	△△△ △△△ △△△ △△△ △△ △△ △△
5	☆ ☆ ☆ ☆ ☆
0	
11	△△△ △△△ △ △△ △△
2	☆ ☆
15	○○○ ○○○ ○○○ ○○ ○○ ○○
1	☆
6	☆ ☆ ☆ ☆ ☆ ☆
20	○○○ ○○○ ○○○ ○○○ ○○ ○○ ○○ ○○

© PolyMath Publishing 2019   May not be reproduced without permission

Kindergarten   4   Understand the relationship between numbers and quantities   ANSWER KEY

Part 2 of 2: Write the number that shows how many objects are in the picture.

Student: _____ ANSWER KEY   _____   _____ / 10 = _____ %

7	○○○○○ ○○
13	□□□□□ □□□□□ □□□
4	✿✿✿✿
12	☆☆☆ ☆☆☆ ☆☆ / ☆☆ ☆☆
8	○○○○ ○○○○
14	✿✿✿✿✿ ✿✿✿✿✿ ✿✿✿✿
19	△△△ △△△ △△△ △△ / △△ △△ △△ △△
17	○○○ ○○○ ○○○ ○ / ○○ ○○ ○○ ○
9	☆☆☆ ☆☆ / ☆☆ ☆☆
16	△△ △△ △△ △△ / △△ △△ △△ △△

© PolyMath Publishing 2019   May not be reproduced without permission

Kindergarten   5   Count how many objects are in a group, up to 10   ANSWER KEY

Tell the student a number from 1 to 10. Have the child draw a group of that many objects, or they can make a group out of classroom objects. Drawings can be of any shape the child is comfortable making.

Student: _____ ANSWER KEY _____   _____ / 10 = _____ %

Number: ___1___   Objects: _____

Number: ___2___   Objects: _____

Number: ___3___   Objects: _____

Number: ___4___   Objects: _____

Number: ___5___   Objects: _____

Number: ___6___   Objects: _____

Number: ___7___   Objects: _____

Number: ___8___   Objects: _____

Number: ___9___   Objects: _____

Number: ___10___   Objects: _____

Kindergarten 6   Compare two groups by matching or counting   ANSWER KEY

Page 1 of 2: In each box, circle the group that has MORE. These two pages may be given to the student at the same time; the assessment is separated to reduce eye strain and visual confusion for the student.

Student: _____ ANSWER KEY _____   _____ / 5 = _____ %

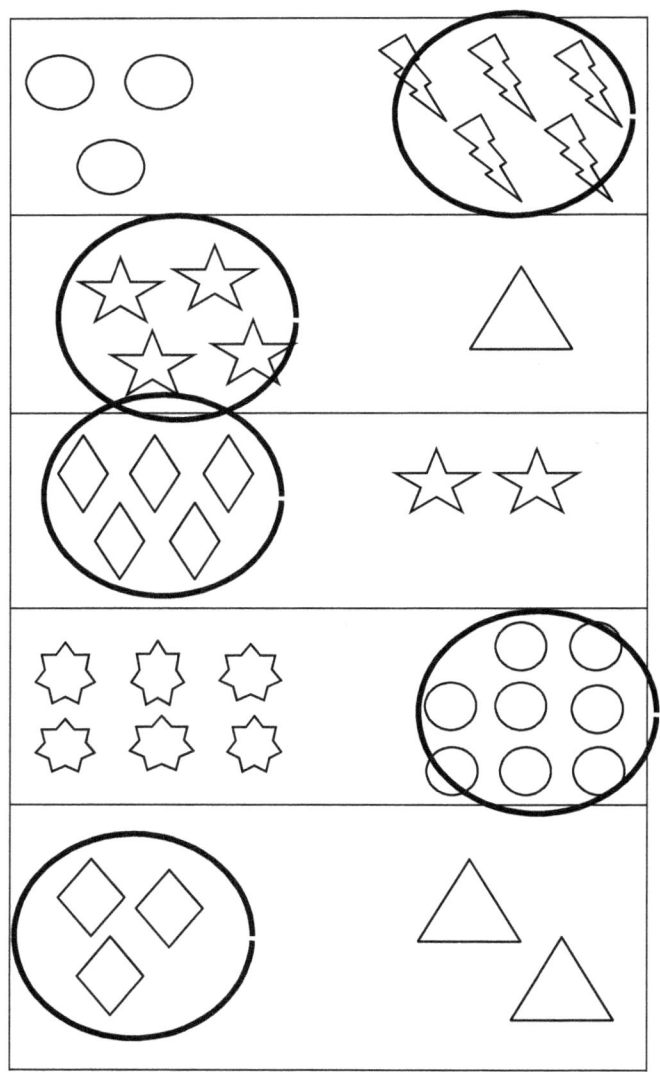

© PolyMath Publishing 2019   May not be reproduced without permission

Kindergarten   6   Compare two groups by matching or counting   ANSWER KEY

Page 2 of 2: In each box, circle the group that has MORE. These two pages may be given to the student at the same time; the assessment is separated to reduce eye strain and visual confusion for the student.

Student: _____ ANSWER KEY   _____   _____ / 5 = _____ %

© PolyMath Publishing 2019   May not be reproduced without permission

Kindergarten   7   Compare two numbers between 1 and 10   ANSWER KEY

In each box, circle the bigger number.

Student: _____ ANSWER KEY   _____   _____ / 12 = _____ %

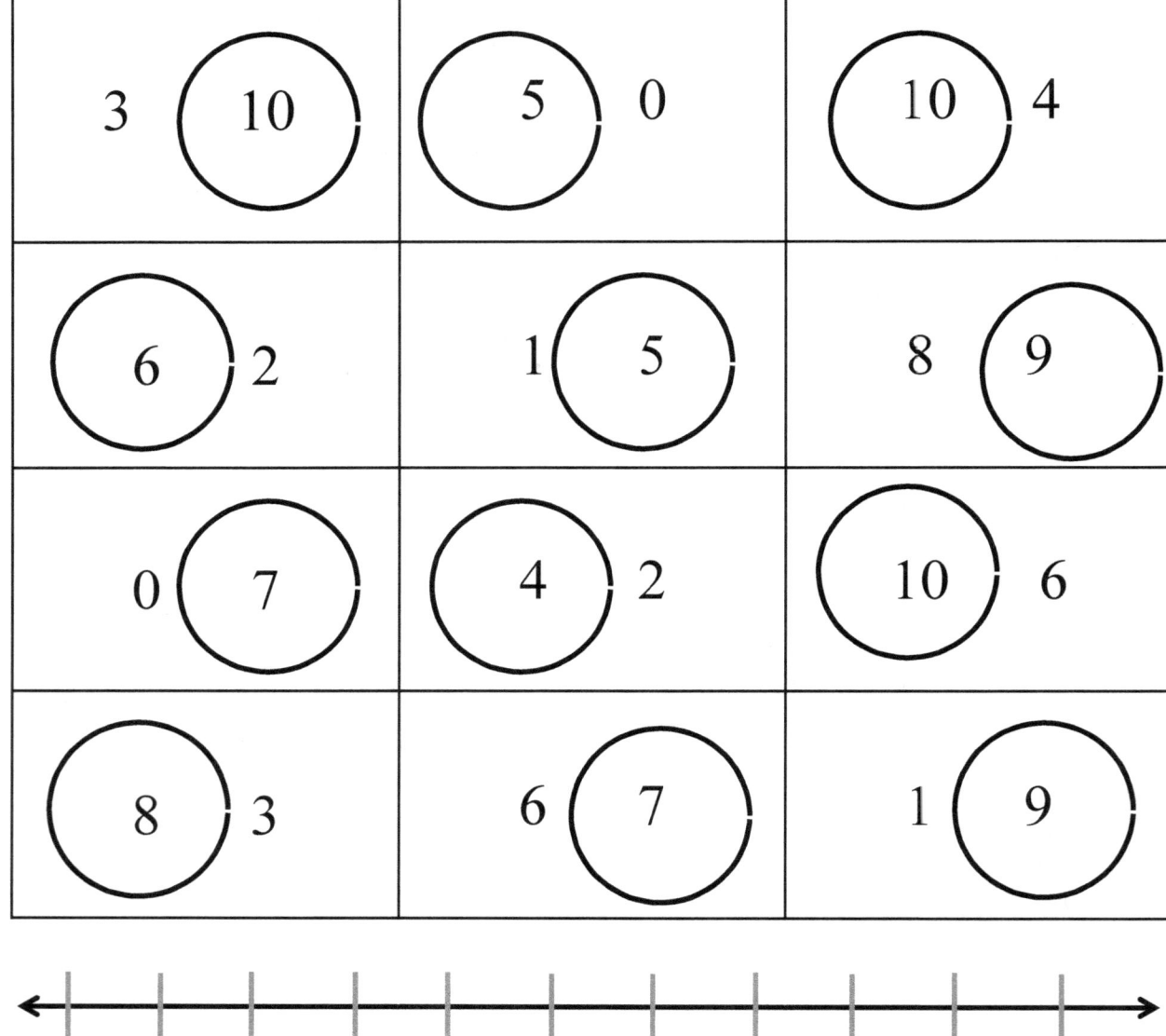

Kindergarten   8   Addition and subtraction word problems   ANSWER KEY

Read each problem to the student. Have them decide if they should add or subtract to solve. Grade one point for the operation (add or subtract) and one point for the answer.

Student: _____ ANSWER KEY _____   _____ / 10 = _____ %

6. James has three apples and Sam has two apples.
   How many apples do they have all together?   3 + 2 = 5 __

7. Luis has three pencils. Nancy has five pencils.
   If they put all their pencils in one basket, how
   many pencils are in the basket?   __ 3 + 5 = 8 ___

8. Sarah has two goldfish. Alex has four goldfish.
   How many fish do they have together?   ___ 2 + 4 = 6 ___

   Draw Sarah's goldfish here:
   Draw Alex's goldfish here:

9. Tommy has ten crayons and gives five
   crayons to Sharon. How many crayons are left?   __ 10 – 5 = 5 ____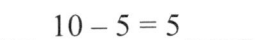

   Draw Tommy's crayons here:

10. If Maria has six apples, and shares two apples with
    her friend Max, how many apples does she have left? _ 6 – 2 = 4 ____

© PolyMath Publishing 2019   May not be reproduced without permission

Kindergarten   9   Decompose numbers less than 10 into addition pairs   ANSWER KEY

Match the number to the addition pair that makes an equal amount. Students may draw a matching line or color matching amounts.

Student: _____ ANSWER KEY   _____   _____ / 10 = _____ %

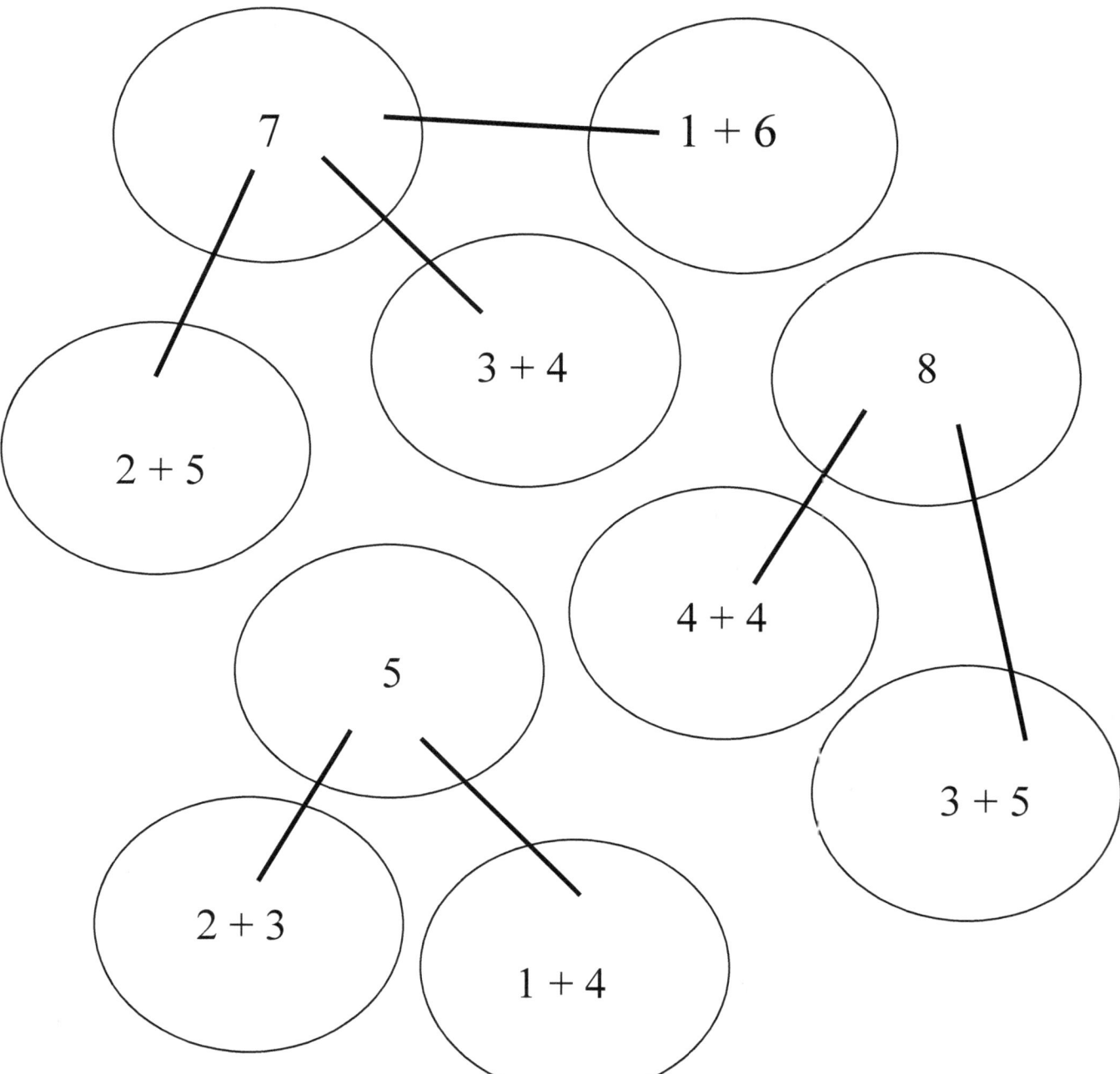

© PolyMath Publishing 2019   May not be reproduced without permission

Kindergarten   10   Use objects or drawings to make ten   ANSWER KEY

Page 1 of 2: Draw a line connecting the boxes to make a set of ten. These two pages may be given to the student at the same time; the assessment is separated to reduce eye strain and visual confusion for the student.

Student: _____ ANSWER KEY   _____ _____ / 5 = _____ %

Example:

Kindergarten   10   Use objects or drawings to make ten   ANSWER KEY

Page 2 of 2: Draw a line connecting the boxes to make a set of ten. These two pages may be given to the student at the same time; the assessment is separated to reduce eye strain and visual confusion for the student.

Student: _____ ANSWER KEY   _____ _____ / 5 = _____ %

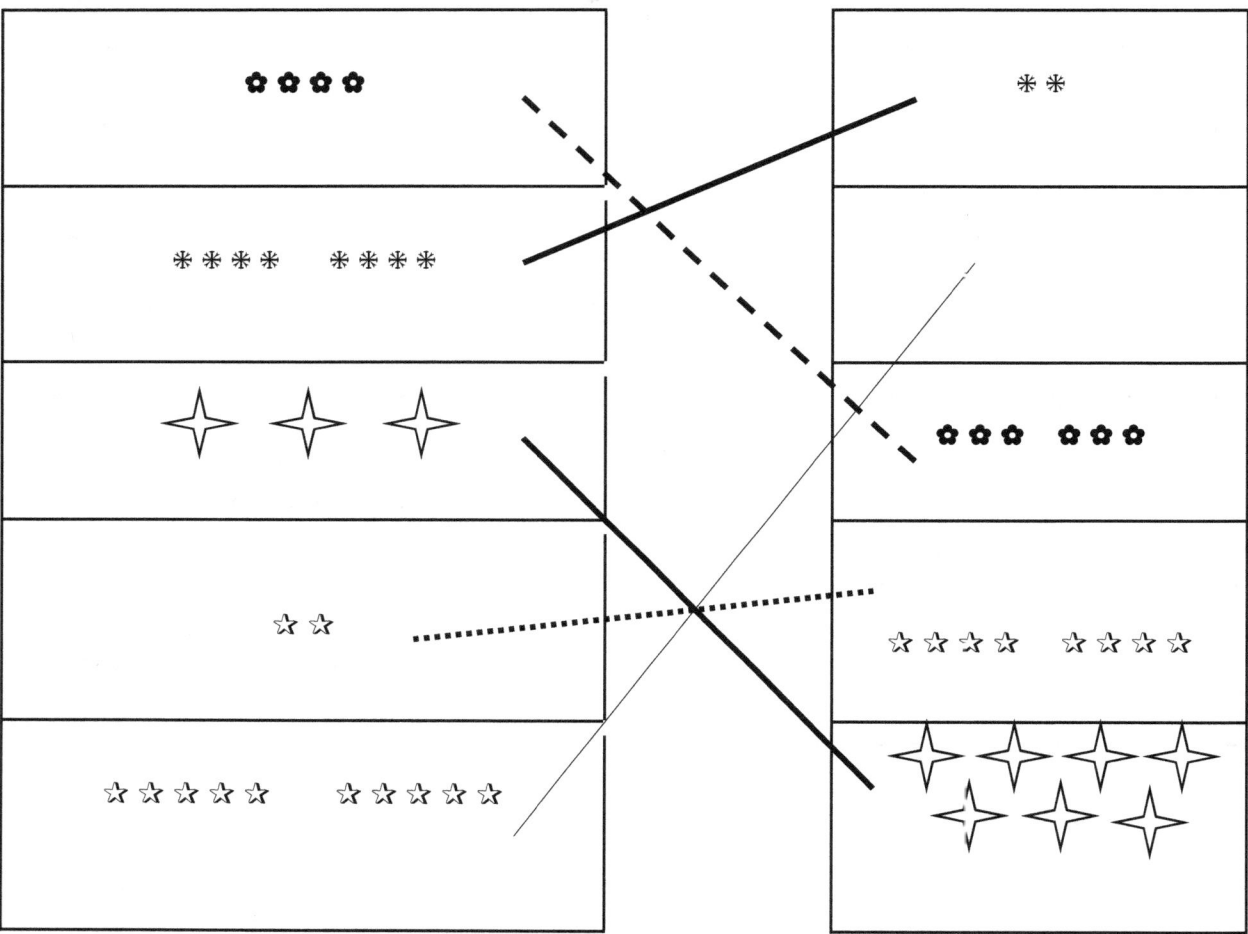

Kindergarten   11   Add and subtract within five   ANSWER KEY

Students with dyscalculia may use support materials, such as a 1-20 chart, number line, or counting on their fingers.

Student: _____ ANSWER KEY   _____   _____ / 20 = _____ %

2 +3 5	4 +1 5	3 +3 6	4 +0 4
5 -3 2	3 -1 2	4 -3 1	5 -0 5
3 +1 4	2 +1 3	2 +2 4	1 +0 1
2 -1 1	4 -1 3	3 -3 0	4 -0 4
1 +3 4	5 -0 5	2 +2 4	3 -0 3

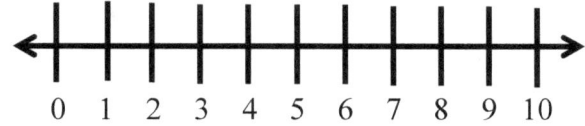

© PolyMath Publishing 2019   May not be reproduced without permission

Kindergarten   12   Decompose 11 through 19 by tens and ones   ANSWER KEY

Write the number in tens and ones, like this: 12 = 10 + 2.

Student: _____ ANSWER KEY   _____   _____/ 10 = _____ %

11	10 + 1
15	10 + 5
18	10 + 8
13	10 + 3
12	10 + 2
16	10 + 6
19	10 + 9
14	10 + 4
17	10 + 7
20	10 + 10

© PolyMath Publishing 2019   May not be reproduced without permission

Kindergarten   13   Weight, length, and height   ANSWER KEY

This is a verbal assessment. Begin with any object in the classroom and ask the student the following series of questions. Record the student's choices with a "yes" or "no" response and a description of the new object.

Student: _____ ANSWER KEY   _____ _____ / 10 = _____ %

Initial object shown to the student: _____ Appropriate answers will relate to the initial object _____

1. Can you find something that is longer than this? _____

2. Can you find something that is shorter than this? _____

3. Can you find something that is taller than this? _____

4. Can you find something that is heavier than this? _____

5. Can you find something that is lighter than this? _____

6. Can you find something that is wider than this? _____

7. Can you find something that is thinner than this? _____

8. Can you find something that is the same height as this? _____

9. Can you find something that is the same weight as this? _____

10. Can you find something that is the same length as this? _____

© PolyMath Publishing 2019   May not be reproduced without permission

Kindergarten   14   Taller and shorter   ANSWER KEY

Student: _____ ANSWER KEY   _____   _____ / 10 = _____ %

Circle the object that is TALLER	Circle the object that is SHORTER
A (circled) A	5 5 (circled)
◇ ◆ (circled)	✸ (circled) ✸
◯ (circled) ○	Z (circled) X
9 (circled) 2	◇ ◆ (circled)
✸ ✸ (circled)	○ ◯ (circled)

© PolyMath Publishing 2019   May not be reproduced without permission

Kindergarten   15   Classify and count with a Venn Diagram   ANSWER KEY

Use the table to fill in the graph.

Student: _____ ANSWER KEY   _____ _____ / 4 = _____ %

Mr. Johnson's class pets:

	Dog	Cat
Jimmy	Yes - Sparky	No
Sam	Yes - Bloo	Yes - RC
Tammy	No	Yes - Fluffy
John	Yes - Jack	Yes - Jill
Lizette	No	No

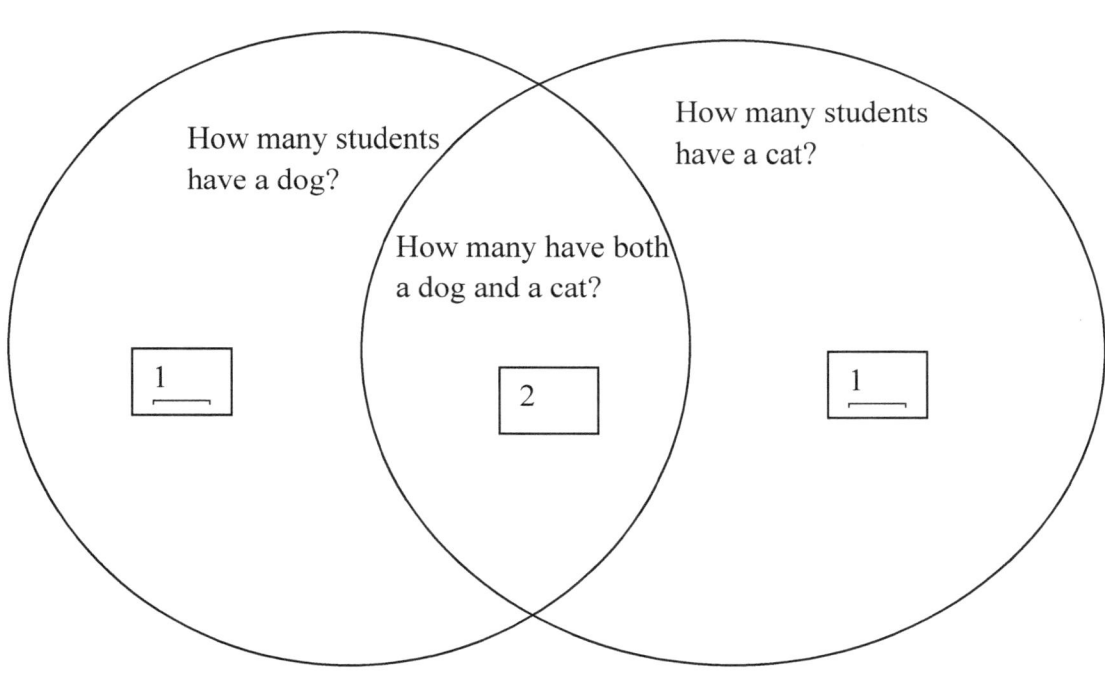

How many students have no pets? _____1____

Kindergarten   16   Describe the Location   ANSWER KEY

Circle the correct answer to complete the sentence.

Student: _____ ANSWER KEY   _____   _____ / 10 = _____ %

The cats are   (INSIDE) / UNDER   the box.

The picture is   NEXT TO   / (ABOVE) the table.

The tree is to the (RIGHT) / LEFT   of the rug.

The table is (ON) / NEXT TO   the rug.

The box is   UNDER   / (ON)   the table.

The rug is (UNDER) / ON   the table.

The picture is to the (LEFT) / RIGHT   of the tree.

The tree is (BEHIND) / IN FRONT OF   the table.

The rug is   BEHIND   / (IN FRONT OF) the tree.

The cats are to the   (RIGHT) / LEFT of the tree.

© PolyMath Publishing 2019   May not be reproduced without permission

Kindergarten 17   Naming shapes   ANSWER KEY

Draw a line matching the word to its shape. Then color all the triangles blue, all the circles red, all the squares green, and all the rectangles yellow.

Student: _____ ANSWER KEY   _____   _____ / 12 = _____ %

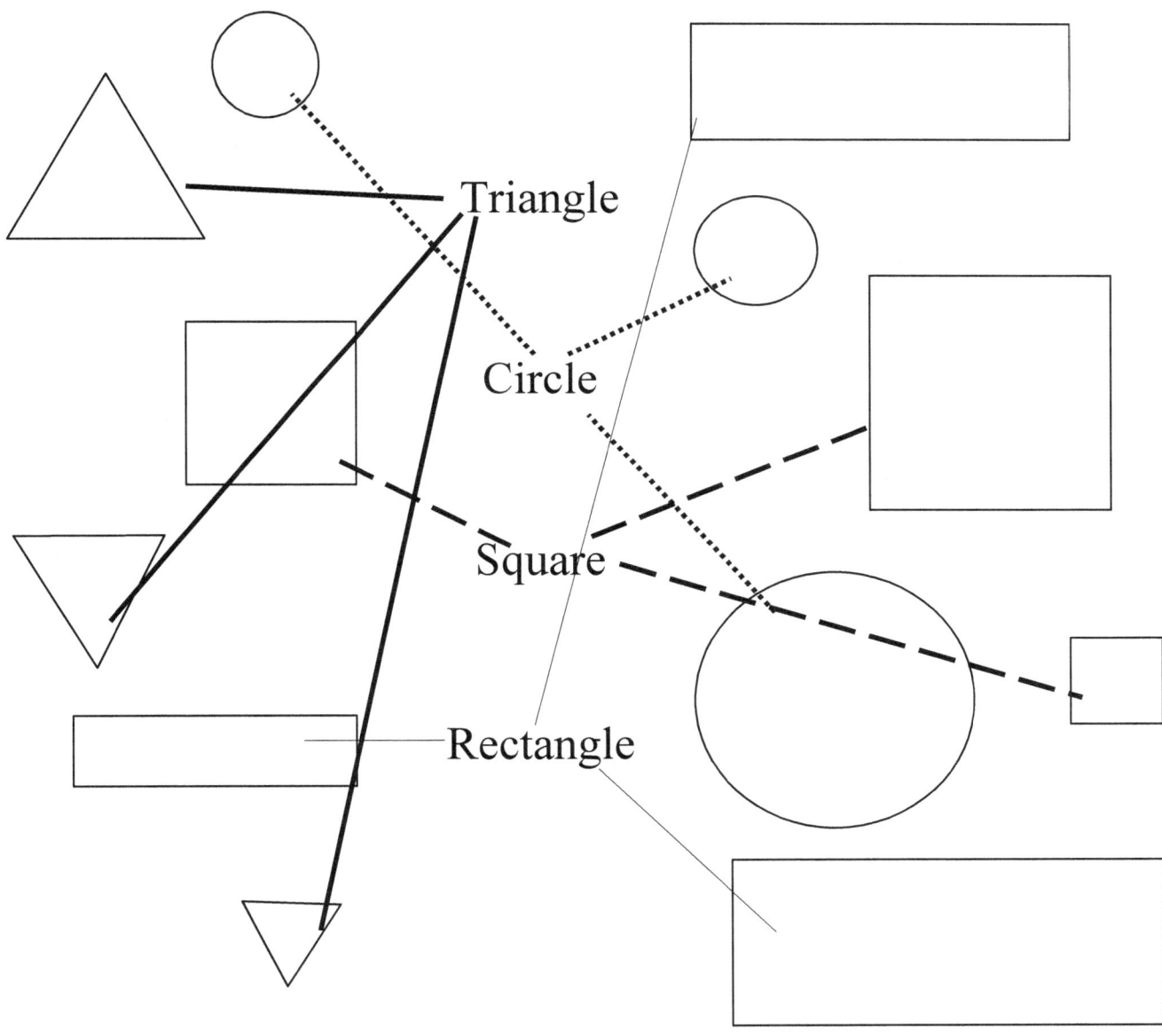

Kindergarten   18   2-D and 3-D shapes   ANSWER KEY

This is an oral assessment. The student can bring an object to you, or they can point out an appropriate object in the room (for example, they might point to a clock or a globe, but not carry one to you). Record the objects the student uses for each shape.

Student: _____ ANSWER KEY   _____   _____ / 8 = _____ %

Can you find a circle? _____ This could be a clock or an eraser tip, or any other circle _____

Can you find a triangle? _____ This could be a triangle in the room, or on a poster _____

Can you find a square? _____ This could be any square object _____

Can you find a sphere? _____ This could be a globe or ball _____

Can you find a cube? _____ This could be a tissue box or container_____

Can you find a prism? _____ This could be a rectangular tissue box or a box of cereal_____

Can you find a rectangle? _____ This could be a shape on a poster _____

Can you find a pyramid? _ This could be an object in the room or a shape on a poster _____

© PolyMath Publishing 2019   May not be reproduced without permission

Kindergarten   19   Same and Different: Shapes   ANSWER KEY

This can be a written or an oral assessment. Give the student a variety of objects: circles, triangles, squares, rectangles, cubes, or any other shape. Let the student describe the ways shapes are similar and different. Ask them about the size, number of corners, flat or solid, or color of the shapes. See if the student can find ten ways they can describe the objects as similar and different.

Student: _____ ANSWER KEY _____ _____ / 10 = _____ %

_____

_____

_____

_____

_____

_____

_____

Possible answers include the number of sides, whether a shape is curved or has straight sides, which one is taller or shorter, or which is heavier or lighter. To score, count the number of ways the student can describe objects in relation to each other, and divide that number by 10.

© PolyMath Publishing 2019   May not be reproduced without permission

Kindergarten   20   Drawing Shapes   ANSWER KEY

Draw each shape and answer the questions about the shape. Teachers can write in student answers to the questions. Drawing the shape counts as one point, and each question counts as one point.

Student: _____ ANSWER KEY _____   _____ / 9 = _____ %

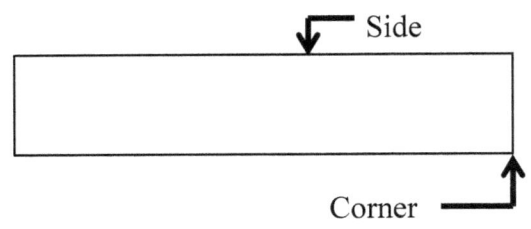

Example: This is a <u>rectangle</u>. It has <u>four</u> sides. It has <u>four</u> corners.

A triangle has _____3_____ sides.

A triangle has _____3_____ corners.

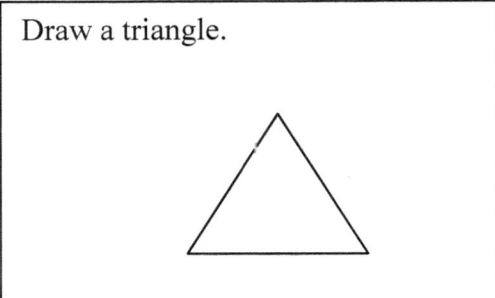

A square has _____4_____ sides.

A square has _____4_____ corners.

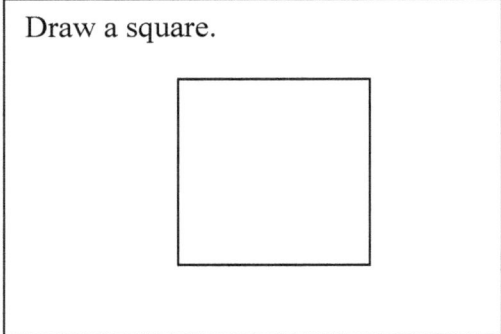

A circle has _____0____ sides.

A circle has _____0____ corners.

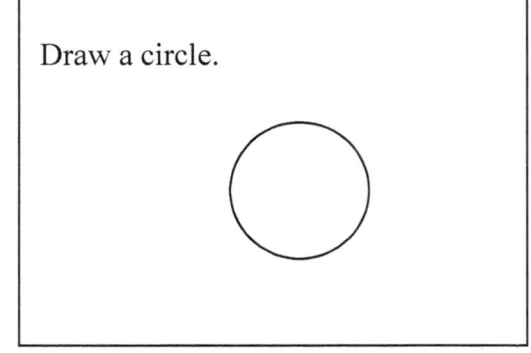

© PolyMath Publishing 2019   May not be reproduced without permission

Kindergarten   21   Use shapes to make new shapes   ANSWER KEY

Have the student use shapes such as triangles, squares, and rectangles to create new shapes. For example, two triangles can make a square or a rectangle. A rectangle and triangle can make a building. The first five shapes are required shapes, and the next five shapes are open to the child's imagination. The student can draw the shapes or use materials and objects in the classroom. Make a note of the materials used.

Student: _____ ANSWER KEY   _____ _____ / 10 = _____ %

Shape: _____ Materials used (drawing, paper cut-outs, objects):

Make a square out of two triangles. _____

Make a rectangle from two triangles. _____

Make a rectangle from two squares. _____

Make a building from squares and triangles. _____

Make a building from rectangles and triangles. _____

Student choice: _____

Student choice: _____

Student choice: _____

To score, record what type of shape the student makes and the materials they use. Each correct shape equals one point out of 10 possible points.

Glossary

This partial glossary explains some of the benefits of PolyMath Assessments and describes some of the students who should use PolyMath Assessments. These assessments and accommodations can be used by every student in any setting, although most students avoid using accommodations they don't need. If a student is showing mastery and wants to work on their own (without a number line, formula sheet, or word bank), by all means, let the student move ahead! If a student is struggling with retention, recall, or information overload, then PolyMath Assessments offers the best means for these students to show what they know.

Dyscalculia: Dyscalculia is a neurological condition that makes it difficult for students to develop numeracy (think of it as "math literacy"). Creating a mental number line, estimating, and recalling formulas can be extremely difficult for these students. PolyMath Assessments are designed specifically for dyscalculia students, because they focus on strengthening the part of the brain where numeracy develops. Accommodations are built into each page and resources are included in each book.

ADD/ADHD: Attention Deficit Disorder and Attention Deficit Hyperactivity Disorder can cause students to lose focus during any activity. This can stop the brain from transferring information from short-term into long-term memory, because the lack of focus means the brain doesn't know which information is worth retaining. PolyMath Assessments have reduced graphics, text, and colors, allowing students to focus on the assessment rather than shifting focus away from critical information.

Processing disorders: Processing speed means the amount of time a person takes to complete a mental task. Some people naturally have a faster processing speed than others, and a slow processing speed is not considered to be a disorder or learning disability on its own. Many students find that the modern classroom pacing goes much too fast for them to keep up with, and this inhibits their learning. Reducing the amount of information presented at one time and reducing anxiety can help these students process information. PolyMath Assessments present information in a relaxed format, building skill and speed slowly, so that all students feel equally able to play the game and enjoy the process.

Dyslexia: Students with dyslexia have trouble processing written language. PolyMath Assessments are oral and visual, allowing these students to excel without the strain of reading. Incorporating these games into a regular classroom-- or even better, into a program of dyslexia-focused interventions and tutoring-- can assist students in strengthening phonics and comprehension through visual, auditory, and tactile centers of the brain, different from the language processing area.

ESOL/ ELL Students: Few things are more impressive, from a cognitive standpoint, than being fluent in more than language. Great job, new English learners! However, becoming fluent while also learning new social skills, cultural norms, and grade-level content in many different subjects can be too much to handle all at one time. PolyMath Assessments support ESOL and ELL learners as they strengthen the neurological connection between information and numerous language databanks.

High-functioning Autism/ Asperger Syndrome: HFA can make it difficult for children to control their emotions and interact with others in a traditional classroom setting. This can stop them from learning as much as they could, at an appropriate grade level, due to missed time in the classroom. PolyMath Assessments reduce the pressure to sit, be quiet, write answers on a worksheet, or answer questions after waiting to be called on by the teacher. This can help HFA students learn more material and demonstrate what they know.

PolyMath Education Services, LLC

PolyMath Education Services offers a complete line of products and services for students who learn differently. Our math assessment workbooks, kindergarten through high school, are written for students with Specific Learning Disorders. The books are easy to use, easy to grade, and easy to store. Each test is designed with built-in accommodations for students who have learning conditions like dyscalculia, dyslexia, ADHD/ADD, processing disorders, and other issues. Tests are clearly marked with appropriate state standards, making them perfect for home school portfolios. PolyMath Assessments can be used with any curriculum or educational program; the workbooks simply offer tests and support materials for students who need accommodations. The accommodations are based on current research and are field tested with students. For traditional classroom teachers, the books offer evaluations that support non-traditional learners in your classroom. For home schoolers, they consolidate your portfolio by showing that your student has met all state requirements for their grade level. For students, they offer the chance to show mastery and enjoy academic success.

The assessments in this book were field-tested with students at the Harvey Homeschool in Tequesta, Florida. The students at Harvey Homeschool have a variety of learning needs, including dyscalculia, dyslexia, high functioning autism, ADHD, and processing disorders. Other students have health-related reasons why a traditional classroom is not right for them, or they home school and enjoy socializing with peers. Students found that the reduced graphics, straight-forward text, and alternative instructions helped them focus on math. You can trust PolyMath Assessments with your child, because we tested PolyMath Assessments with ours first. Find out more at www.polymathpublishing.net.

INDEX OF STATE STANDARDS

	Alabama	Alaska	Arizona	Arkansas: Add AR.Math.Content to front	California	Colorado	Connecticut: Add CCSS.Math.Content to front
#1, Page 6	K-CC1	K.CC1	K.CC.A.1	K.CC.A.1	K-CC1	1.1.a.i	K.CC.A.1
#2, Page 8	K-CC2	K.CC2	K.CC.A.2	K.CC.A.2	K-CC2	1.1.a.ii	K.CC.A.2
#3, Page 10	K-CC3	K.CC3	K.CC.A.3	K.CC.A.3	K-CC3	1.1.a.iii	K.CC.A.3
#4, Page 16	K-CC4	K.CC4	K.CC.B.4	K.CC.B.4	K-CC4	1.1.b.i	K.CC.B.4
#5, Page 20	K-CC5	K.CC5	K.CC.B.5	K.CC.B.5	K-CC5	1.1.b.ii	K.CC.B.5
#6, Page 22	K-CC6	K.CC6	K.CC.C.6	K.CC.C.6	K-CC6	1.1.c.i & iii	K.CC.C.6
#7, Page 26	K-CC7	K.CC7	K.CC.C.7	K.CC.C.7	K-CC7	1.1.c.ii	K.CC.C.7
#8, Page 28	K-OA1/2	K.OA1/2	K.OA.A.1/2	K.OA.A.1/2	K-OA1/2	1.2.a.ii	K.OA.A.1/2
#9, Page 30	K-OA3	K.OA3	K.OA.A.3	K.OA.A.3	K-OA3	1.2.a.iii	K.OA.A.3
#10, Page 32	K-OA4	K.OA4	K.OA.A.4	K.OA.A.4	K-OA4	1.2.a.iv	K.OA.A.4
#11, Page 36	K-OA5	K.OA5	K.OA.A.5	K.OA.A.5	K-OA5	1.2.b	K.OA.A.5
#12, Page 38	K-NBT-1	K.NBT1	K.NBT.A.1	K.NBT.A.1	K-NBT-1	1.2.c	K.NBT.A.1
#13, Page 40	K-MD1	K.MD1	K.MD.A.1	K.MD.A.1	K-MD1	4.2.a.i	K.MD.A.1
#14, Page 42	K-MD2	K.MD2	K.MD.A.2	K.MD.A.2	K-MD2	4.2.a.ii	K.MD.A.2
#15, Page 44	K-MD3	K.MD3	K.MD.B.3	K.MD.B.3	K-MD3	4.2.b	K.MD.B.3
#16, Page 46	K-G1	K.G1	K.G.A.1	K.G.A.1	K-G1	Not Applicable	K.G.A.1
#17, Page 48	K-G2	K.G2	K.G.A.2	K.G.A.2	K-G2	4.1.a.i & ii	K.G.A.2
#18, Page 50	K-G3	K.G3	K.G.A.3	K.G.A.3	K-G3	4.1.a.iii	K.G.A.3
#19, Page 52	K-G4	K.G4	K.G.B.4	K.G.B.4	K-G4	4.1.b.i-iii	K.G.B.4
#20, Page 54	K-G5	K.G5	K.G.B.5	K.G.B.5	K-G5	Not Applicable	K.G.B.5
#21, Page 56	K-G6	K.G6	K.G.B.6	K.G.B.6	K-G6	Not Applicable	K.G.B.6

INDEX OF STATE STANDARDS

	Delaware	Florida	Georgia	Hawaii	Idaho	Illinois	Indiana
#1, Page 6	K.CC.A.1	MAFS.K.CC.1.1	MGSEK.CC.1	K.CC.A.1	K.CC.A.1	K.CC.A.1	K.NS.1
#2, Page 8	K.CC.A.2	MAFS.K.CC.1.2	MGSEK.CC.2	K.CC.A.2	K.CC.A.2	K.CC.A.2	K.NS.4
#3, Page 10	K.CC.A.3	MAFS.K.CC.1.3	MGSEK.CC.3	K.CC.A.3	K.CC.A.3	K.CC.A.3	K.NS.2
#4, Page 16	K.CC.B.4	MAFS.K.CC.2.4	MGSEK.CC.4	K.CC.B.4	K.CC.B.4	K.CC.B.4	K.NS.2
#5, Page 20	K.CC.B.5	MAFS.K.CC.2.5	MGSEK.CC.5	K.CC.B.5	K.CC.B.5	K.CC.B.5	K.NS.2
#6, Page 22	K.CC.C.6	MAFS.K.CC.3.6	MGSEK.CC.6	K.CC.C.6	K.CC.C.6	K.CC.C.6	K.NS.8
#7, Page 26	K.CC.C.7	MAFS.K.CC.3.7	MGSEK.CC.7	K.CC.C.7	K.CC.C.7	K.CC.C.7	K.NS.8
#8, Page 28	K.OA.A.1/2	MAFS.K.OA.1.1/2	MGSEK.OA.1/2	K.OA.A.1/2	K.OA.A.1/2	K.OA.A.1/2	K.CA.2
#9, Page 30	K.OA.A.3	Not Applicable	MGSEK.OA.3	K.OA.A.3	K.OA.A.3	K.OA.A.3	K.CA.4
#10, Page 32	K.OA.A.4	MAFS.K.OA.1.4	MGSEK.OA.4	K.OA.A.4	K.OA.A.4	K.OA.A.4	K.CA.4
#11, Page 36	K.OA.A.5	MAFS.K.OA.1.5	MGSEK.OA.5	K.OA.A.5	K.OA.A.5	K.OA.A.5	K.CA.2
#12, Page 38	K.NBT.A.1	MAFS.K.NBT.1.1	MGSEK.NBT.1	K.NBT.A.1	K.NBT.A.1	K.NBT.A.1	K.CA.3
#13, Page 40	K.MD.A.1	MAFS.K.MD.1.1	MGSEK.MD.1	K.MD.A.1	K.MD.A.1	K.MD.A.1	K.MD.1
#14, Page 42	K.MD.A.2	MAFS.K.MD.1.2	MGSEK.MD.2	K.MD.A.2	K.MD.A.2	K.MD.A.2	K.MD.2
#15, Page 44	K.MD.B.3	MAFS.K.MD.2.3	MGSEK.MD.3	K.MD.B.3	K.MD.B.3	K.MD.B.3	K.DA.1
#16, Page 46	K.G.A.1	MAFS.K.G.1.1	MGSEK.G.A.1	K.G.A.1	K.G.A.1	K.G.A.1	K.G.1
#17, Page 48	K.G.A.2	MAFS.K.G.1.2	MGSEK.G.A.2	K.G.A.2	K.G.A.2	K.G.A.2	Not Applicable
#18, Page 50	K.G.A.3	MAFS.K.G.1.3	MGSEK.G.A.3	K.G.A.3	K.G.A.3	K.G.A.3	K.G.2
#19, Page 52	K.G.B.4	MAFS.K.G.2.4	MGSEK.G.B.4	K.G.B.4	K.G.B.4	K.G.B.4	Not Applicable
#20, Page 54	K.G.B.5	MAFS.K.G.2.5	MGSEK.G.B.5	K.G.B.5	K.G.B.5	K.G.B.5	K.G.3
#21, Page 56	K.G.B.6	MAFS..K.G.2.6	MGSEK.G.B.6	K.G.B.6	K.G.B.6	K.G.B.6	K.G.4

© PolyMath Publishing 2019

May not be reproducec without permission

INDEX OF STATE STANDARDS

	Iowa	Kansas	Kentucky	Louisiana	Maine	Maryland	Massachusetts
#1, Page 6	K.CC.A.1	K.CC.1	K.CC.1	K.CC.A.1	K.CC.A.1	K.CC.A.1	K.CC.A.1
#2, Page 8	K.CC.A.2	K.CC.2	K.CC.2	K.CC.A.2	K.CC.A.2	K.CC.A.2	K.CC.A.2
#3, Page 10	K.CC.A.3	K.CC.3	K.CC.3	K.CC.A.3	K.CC.A.3	K.CC.A.3	K.CC.A.3
#4, Page 16	K.CC.B.4	K.CC.4	K.CC.4	K.CC.B.4	K.CC.B.4	K.CC.B.4	K.CC.B.4
#5, Page 20	K.CC.B.5	K.CC.5	K.CC.5	K.CC.B.5	K.CC.B.5	K.CC.B.5	K.CC.B.5
#6, Page 22	K.CC.C.6	K.CC.6	K.CC.6	K.CC.C.6	K.CC.C.6	K.CC.C.6	K.CC.C.6
#7, Page 26	K.CC.C.7	K.CC.7	K.CC.7	K.CC.C.7	K.CC.C.7	K.CC.C.7	K.CC.C.7
#8, Page 28	K.OA.A.1/2	K.OA.1/2	K.OA.1/2	K.OA.A.1/2	K.OA.A.1/2	K.OA.A.1/2	K.OA.A.1/2
#9, Page 30	K.OA.A.3	K.OA.3	K.OA.3	K.OA.A.3	K.OA.A.3	K.OA.A.3	K.OA.A.3
#10, Page 32	K.OA.A.4	K.OA.4	K.OA.4	K.OA.A.4	K.OA.A.4	K.OA.A.4	K.OA.A.4
#11, Page 36	K.OA.A.5	K.OA.5	K.OA.5	K.OA.A.5	K.OA.A.5	K.OA.A.5	K.OA.A.5
#12, Page 38	K.NBT.A.1	K.NBT.1	K.NBT.1	K.NBT.A.1	K.NBT.A.1	K.NBT.A.1	K.NBT.A.1
#13, Page 40	K.MD.A.1	K.MD.1	K.MD.1	K.MD.A.1	K.MD.A.1	K.MD.A.1	K.MD.A.1
#14, Page 42	K.MD.A.2	K.MD.2	K.MD.2	K.MD.A.2	K.MD.A.2	K.MD.A.2	K.MD.A.2
#15, Page 44	K.MD.B.3	K.MD.3	K.MD.3	K.MD.B.3	K.MD.B.3	K.MD.B.3	K.MD.B.3
#16, Page 46	K.G.A.1	K.G.1	K.G.1	K.G.A.1	K.G.A.1	K.G.A.1	K.G.A.1
#17, Page 48	K.G.A.2	K.G.2	K.G.2	K.G.A.2	K.G.A.2	K.G.A.2	K.G.A.2
#18, Page 50	K.G.A.3	K.G.3	K.G.3	K.G.A.3	K.G.A.3	K.G.A.3	K.G.A.3
#19, Page 52	K.G.B.4	K.G.4	K.G.4	K.G.B.4	K.G.B.4	K.G.B.4	K.G.B.4
#20, Page 54	K.G.B.5	K.G.5	K.G.5	K.G.B.5	K.G.B.5	K.G.B.5	K.G.B.5
#21, Page 56	K.G.B.6	K.G.6	K.G.6	K.G.B.6	K.G.B.6	K.G.B.6	K.G.B.6

INDEX OF STATE STANDARDS

	Michigan	Minnesota	Mississippi	Missouri	Montana	Nebraska	Nevada
#1, Page 6	K.CC.1	K.1.1.2	K.CC.1	NSA1	K.CC.1	MA 0.1.1.a	K.CC.A.1
#2, Page 8	K.CC.2	Not Applicable	K.CC.2	NSA2	K.CC.2	MA 0.1.1.a	K.CC.A.2
#3, Page 10	K.CC.3	K.1.1.2	K.CC.3	NSA4	K.CC.3	MA 0.1.1.c	K.CC.A.3
#4, Page 16	K.CC.4	K.1.1.3	K.CC.4	NSB5	K.CC.4	MA 0.1.1.f	K.CC.B.4
#5, Page 20	K.CC.5	K.1.1.5	K.CC.5	NSB6/7	K.CC.5	MA 0.1.1.e	K.CC.B.5
#6, Page 22	K.CC.6	K.1.2.1	K.CC.6	NSB8/9	K.CC.6	MA 0.1.1.h	K.CC.C.6
#7, Page 26	K.CC.7	K.1.2.2	K.CC.7	NSB10	K.CC.7	MA 0.1.2.a	K.CC.C.7
#8, Page 28	K.OA.1/2	K.1.1.2	K.OA.1/2	RAA1	K.OA.1/2	MA 0.2.1.a	K.OA.A.1/2
#9, Page 30	K.OA.3	Not Applicable	K.OA.3	RAA2	K.OA.3	MA 0.2.1.b	K.OA.A.3
#10, Page 32	K.OA.4	Not Applicable	K.OA.4	RAA3/4	K.OA.4	MA 0.2.3.a	K.OA.A.4
#11, Page 36	K.OA.5	Not Applicable	K.OA.5	RAA2	K.OA.5	MA 0.1.1.g	K.OA.A.5
#12, Page 38	K.NBT.1	K.3.1.2	K.NBT.1	NBTA1	K.NBT.1	MA 0.3.3.a	K.NBT.A.1
#13, Page 40	K.MD.1	K.3.2.1	K.MD.1	GMA1	K.MD.1	MA 0.4.2.a	K.MD.A.1
#14, Page 42	K.MD.2	Not Applicable	K.MD.2	GMA2	K.MD.2	MA 0.3.2.a	K.MD.A.2
#15, Page 44	K.MD.3	K.3.1.1	K.MD.3	DSA1/2	K.MD.3	MA 0.31.a	K.MD.B.3
#16, Page 46	K.G.1	K.3.2.2	K.G.1	GMC7	K.G.1	MA 0.3.1.b	K.G.A.1
#17, Page 48	K.G.2	K.3.1.1	K.G.2	GMC6	K.G.2	MA 0.3.1.d	K.G.A.2
#18, Page 50	K.G.3	K.3.2.2	K.G.3	GMC6	K.G.3	MA 0.3.1.d	K.G.A.3
#19, Page 52	K.G.4	K.3.1.1	K.G.4	GMC8	K.G.4	MA 0.3.1.b	K.G.B.4
#20, Page 54	K.G.5	K.3.2.2	K.G.5	GMC9	K.G.5	MA 0.3.1.b	K.G.B.5
#21, Page 56	K.G.6	Not Applicable	K.G.6	GMC10	K.G.6	MA 0.3.1.e	K.G.B.6

INDEX OF STATE STANDARDS

	New Hampshire	New Jersey	New Mexico	New York	North Carolina	North Dakota	Ohio
#1, Page 6	K.CC.1	K.CC.A.1	K.CC.A.1	K.CC.A.1	NC.K.CC.A.1	K.CC.1	K.CC.1
#2, Page 8	K.CC.2	K.CC.A.2	K.CC.A.2	K.CC.A.2	NC.K.CC.A.2	K.CC.2	K.CC.2
#3, Page 10	K.CC.3	K.CC.A.3	K.CC.A.3	K.CC.A.3	NC.K.CC.A.3	K.CC.3	K.CC.3
#4, Page 16	K.CC.4	K.CC.B.4	K.CC.B.4	K.CC.B.4	NC.K.CC.B.4	K.CC.4	K.CC.4
#5, Page 20	Not Applicable	K.CC.B.5	K.CC.B.5	K.CC.B.5	NC.K.CC.B.5	Not Applicable	Not Applicable
#6, Page 22	K.CC.6	K.CC.C.6	K.CC.C.6	K.CC.C.6	NC.K.CC.C.6	K.CC.6	K.CC.6
#7, Page 26	K.CC.7	K.CC.C.7	K.CC.C.7	K.CC.C.7	NC.K.CC.C.7	K.CC.7	K.CC.7
#8, Page 28	K.OA.1/2	K.OA.A.1/2	K.OA.A.1/2	K.OA.A.1/2	NC.K.OA.A.1/2	K.OA.1/2	K.OA.1/2
#9, Page 30	K.OA.3	K.OA.A.3	K.OA.A.3	K.OA.A.3	NC.K.OA.A.3	K.OA.3	K.OA.3
#10, Page 32	K.OA.4	K.OA.A.4	K.OA.A.4	K.OA.A.4	NC.K.OA.A.4	K.OA.4	K.OA.4
#11, Page 36	K.OA.5	K.OA.A.5	K.OA.A.5	K.OA.A.5	NC.K.OA.A.5	K.OA.5	K.OA.5
#12, Page 38	K.NBT.1	K.NBT.A.1	K.NBT.A.1	K.NBT.A.1	NC.K.NBT.A.1	K.NBT.1	K.NBT.1
#13, Page 40	K.MD.1	K.MD.A.1	K.MD.A.1	K.MD.A.1	NC.K.MD.A.1	K.MD.1	K.MD.1
#14, Page 42	K.MD.2	K.MD.A.2	K.MD.A.2	K.MD.A.2	NC.K.MD.A.2	K.MD.2	K.MD.2
#15, Page 44	K.MD.3	K.MD.B.3	K.MD.B.3	K.MD.B.3	NC.K.MD.B.3	K.MD.3	K.MD.3
#16, Page 46	K.G.1	K.G.A.1	K.G.A.1	K.G.A.1	NC.K.G.A.1	K.G.1	K.G.1
#17, Page 48	K.G.2	K.G.A.2	K.G.A.2	K.G.A.2	NC.K.G.A.2	K.G.2	K.G.2
#18, Page 50	K.G.3	K.G.A.3	K.G.A.3	K.G.A.3	NC.K.G.A.3	K.G.3	K.G.3
#19, Page 52	K.G.4	K.G.B.4	K.G.B.4	K.G.B.4	NC.K.G.B.4	K.G.4	K.G.4
#20, Page 54	K.G.5	K.G.B.5	K.G.B.5	K.G.B.5	NC.K.G.B.5	K.G.5	K.G.5
#21, Page 56	K.G.6	K.G.B.6	K.G.B.6	K.G.B.6	NC.K.G.B.6	K.G.6	K.G.6

INDEX OF STATE STANDARDS

	Oklahoma	Oregon	Pennsylvania	Rhode Island	South Carolina	South Dakota	Tennessee
#1, Page 6	K.N.1.1	K.CC.1	CC.2.1.K.A.1	K.CC.1	K.NS.1	K.CC.1	K.CC.A.1
#2, Page 8	K.N.1.3	K.CC.2	CC.2.1.K.A.1	K.CC.2	K.NS.2	K.CC.2	K.CC.A.2
#3, Page 10	K.N.1.2	K.CC.3	CC.2.1.K.A.1	K.CC.3	K.NS.3	K.CC.3	K.CC.A.3
#4, Page 16	K.N.1.5	K.CC.4	CC.2.1.K.A.2	K.CC.4	K.NS.4	K.CC.4a	K.CC.B.4
#5, Page 20	K.N.1.6	Not Applicable	CC.2.1.K.A.2	K.CC.5	K.NS.5	K.CC.5	K.CC.B.5
#6, Page 22	K.N.1.8	K.CC.6	CC.2.1.K.A.3	K.CC.6	K.NS.7	K.CC.4b	K.CC.C.6
#7, Page 26	K.N.1.8	K.CC.7	CC.2.1.K.A.3	K.CC.7	K.NS.8	K.CC.6/7	K.CC.C.7
#8, Page 28	K.N.2.1	K.OA.1/2	CC.2.2.K.A.1	K.OA.1/2	K.ATO.1	K.OA.1/2	K.OA.A.1/2
#9, Page 30	K.N.2.1	K.OA.3	CC.2.2.K.A.1	K.OA.3	K.ATO.1	K.OA.3	K.OA.A.3
#10, Page 32	K.N.2.1	K.OA.4	CC.2.2.K.A.1	K.OA.4	K.ATO.2/3	K.OA.4	K.OA.A.4
#11, Page 36	K.N.2.1	K.OA.5	CC.2.2.K.A.1	K.OA.5	K.ATO.5	K.OA.5	K.OA.A.5
#12, Page 38	Not Applicable	K.NBT.1	CC.2.1.K.B.1	K.NBT.1	K.NSBT.1	K.NBT.1	K.NBT.A.1
#13, Page 40	K.GM.2.1	K.MD.1	CC.2.4.K.A.1	K.MD.1	K.MDA.1	K.MD.1	K.MD.A.1
#14, Page 42	K.GM.1.2	K.MD.2	CC.2.4.K.A.1	K.MD.2	K.MDA.1	K.MD.2	K.MD.A.2
#15, Page 44	K.D.1	K.MD.3	CC.2.4.K.A.4	K.MD.3	K.MDA.4	K.MD.3	K.MD.B.3
#16, Page 46	K.GM.2.1	K.G.1	CC.2.3.K.A.2	K.G.1	K.G.1	K.G.1	K.G.A.1
#17, Page 48	K.GM.1.1	K.G.2	CC.2.3.K.A.1	K.G.2	K.G.2	K.G.2	K.G.A.2
#18, Page 50	K.GM.1.1	K.G.3	CC.2.3.K.A.1	K.G.3	K.G.3/4	K.G.3	K.G.A.3
#19, Page 52	K.GM.1.2	K.G.4	CC.2.3.K.A.2	K.G.4	K.MDA.2	K.G.4	K.G.B.4
#20, Page 54	K.GM.1.5	K.G.5	CC.2.3.K.A.2	K.G.5	K.G.5	K.G.5	K.G.B.5
#21, Page 56	K.GM.1.4	K.G.6	CC.2.3.K.A.2	K.G.6	K.G.6	K.G.6	K.G.B.6

INDEX OF STATE STANDARDS

	Texas	Utah	Vermont	Virginia	Washington	West Virginia	Wisconsin	Wyoming
#1, Page 6	2A	K.CC.1	K.CC.1	K.3	K.CC.1	M.K.1	K.CC.1	K.CC.A.1
#2, Page 8	2A	K.CC.2	K.CC.2	K.3	K.CC.2	M.K.2	K.CC.2	K.CC.A.2
#3, Page 10	2B	K.CC.3	K.CC.3	K.1	K.CC.3	M.K.3	K.CC.3	K.CC.A.3
#4, Page 16	2C	K.CC.4	K.CC.4	K.1	K.CC.4	M.K.4	K.CC.4	K.CC.B.4
#5, Page 20	2D	Not Applicable	Not Applicable	K.2	Not Applicable	M.K.5	Not Applicable	K.CC.B.5
#6, Page 22	2G/I	K.CC.6	K.CC.6	K.2	K.CC.6	M.K.6	K.CC.6	K.CC.C.6
#7, Page 26	2H	K.CC.7	K.CC.7	K.2	K.CC.7	M.K.7	K.CC.7	K.CC.C.7
#8, Page 28	2H	K.OA.1/2	K.OA.1/2	K.6	K.OA.1/2	M.K.9	K.OA.1/2	K.OA.A.1/2
#9, Page 30	3A	K.OA.3	K.OA.3	K.4	K.OA.3	M.K.10	K.OA.3	K.OA.A.3
#10, Page 32	3B	K.OA.4	K.OA.4	K.6	K.OA.4	M.K.11	K.OA.4	K.OA.A.4
#11, Page 36	3B	K.OA.5	K.OA.5	K.6	K.OA.5	M.K.12	K.OA.5	K.OA.A.5
#12, Page 38	3B	K.NBT.1	K.NBT.1	K.6	K.NBT.1	M.K.13	K.NBT.1	K.NBT.A.1
#13, Page 40	7A	K.MD.1	K.MD.1	K.9	K.MD.1	M.K.14	K.MD.1	K.MD.A.1
#14, Page 42	7B	K.MD.2	K.MD.2	K.9	K.MD.2	M.K.15	K.MD.2	K.MD.A.2
#15, Page 44	8B	K.MD.3	K.MD.3	K.11	K.MD.3	M.K.16	K.MD.3	K.MD.B.3
#16, Page 46	Not Applicable	K.G.1	K.G.1	K.10	K.G.1	M.K.17	K.G.1	K.G.A.1
#17, Page 48	6A	K.G.2	K.G.2	K.10	K.G.2	M.K.18	K.G.2	K.G.A.2
#18, Page 50	6B	K.G.3	K.G.3	K.10	K.G.3	M.K.19	K.G.3	K.G.A.3
#19, Page 52	6D	K.G.4	K.G.4	K.12	K.G.4	M.K.20	K.G.4	K.G.B.4
#20, Page 54	6F	K.G.5	K.G.5	K.10	K.G.5	M.K.21	K.G.5	K.G.B.5
#21, Page 56	6F	K.G.6	K.G.6	K.10	K.G.6	M.K.22	K.G.6	K.G.B.6

Honora Wall, Author

Honora Wall is an educator, author, and founder of PolyMath Publishing. She holds a Master's in Applied Curriculum & Instruction from the University of Central Florida and is earning her Ed.D. in Education, Curriculum & Instruction, from Concordia University Chicago. She taught elementary and middle school as a classroom teacher. Through her company, What the (f)unction, she worked as a private tutor focusing on high school and college students. Through this expansive background, she realized how small changes in instruction and assessment could have a big impact on student growth. This led her to develop PolyMath Assessments and a training and curriculum program that supports students with a variety of learning issues. She hopes that her work will transform both student's experiences with math and their self-confidence in their learning abilities. Please send your comments and suggestions to her through www.polymathpublishing.net.

Chris Wall, Illustrator

Chris Wall is the resident artist at What the (f)unction, Inc. and PolyMath Publishing. He creates his own artwork through his brand, BadlyDrawnRobot. His experience includes a variety of positions working on film, television, and commercial projects in Lake Worth, Florida, Orlando, Florida, and Chicago, Illinois. He studied at F.I.R.S.T. Institute in Orlando and Columbia College Chicago.